Isaac Newton

of Woolsthorpe Manor and
The King's School, Grantham

by

the Year 7 Boys of the School
with John Haden

First published by Barny Books,
All rights reserved
Copyright © John Haden 2010

No part of this publication may be reproduced or transmitted in any way or by any means, including electronic storage and retrieval, without prior permission of the publisher.

ISBN No: 978.1.906542.33.7
Publishers: Barny Books
              Hough on the Hill,
              Grantham,
              Lincolnshire
              NG32 2BB

              Tel: 01400 250246

Copies of this book may be obtained from:

              The King's School,
              Brook Street,
              Grantham,
              Lincolnshire, NG31 6RP

              Tel: 01476 563180

1. 350 Years Ago

Three hundred and fifty years ago, a young farmer from Woolsthorpe returned to The King's School, Grantham. His name was Isaac Newton and for about a year, he had been at home. He failed to become a successful farmer. He had lost a horse, allowed his sheep to escape into other people's crops and so annoyed his mother that she eventually agreed to send him back to school.

In the same year, a group of leading thinkers in London got together and formed a discussion group. They asked the King, Charles II, for his permission to call their group the Royal Society. They chose as a Latin motto: 'Nullius in verba' or 'take no-one's word for it' or even 'check it out for yourself'. So this year is the 350th anniversary of the founding of the Royal Society. In writing Isaac Newton's story, we have tried to follow their motto in researching the life and achievements of England's greatest scientist.   (by Tom H. and Ben M.)

2. Successful Farmers

The Newton menfolk were yeoman farmers, making good money from the backs of their Lincoln Longwool sheep. For six generations in the 16th, 17th and early 18th centuries, they owned and farmed land in and around the small Lincolnshire village of Woolsthorpe by Colsterworth, seven miles south of the market town of Grantham. By 1623, Robert Newton had made enough money to buy the Manor House with the barns and land around it, adding them to his existing fifty acres in the village. A house had stood on the site for about three hundred years but Robert decided to rebuild it in warm Ancaster limestone with mullioned windows on the west front and a row of dormers set in a thatched roof. It was a fine house for a man with social ambitions and with the property came the title Lord of the Manor.

Woolsthorpe Manor in the C17th from William Stukeley's sketch

Robert decided to give the house as a wedding dowry to his son, Isaac. When Isaac married Hannah Ayscough in April 1642, the old man must have been quietly satisfied. For the Ayscoughs were 'gentry', a family well-known all over Lincolnshire as people of property and standing, rich enough to send their sons to school and university. They even provided their daughters with a smattering of 'book learning', so Hannah could read and write. Not well, but enough to send letters and to mark her out as different from the illiterate people of the village. Few had any use for books in Woolsthorpe.

Isaac's father was called Isaac as well. He couldn't read or write but this didn't matter very much as he was a farmer. On his bedroom wall at the Manor House, he made a farmer's tally to show how many animals were in each field on the farm, a diagram with dots for each animal. He

was a very wealthy farmer and, when he died, he had 246 sheep and barns full of wheat, barley and hay. To get the crops in before they were destroyed by bad weather, the local people and any travellers would come to help on the farm, because it was critical for them to have food for the winter. In return they got food and accommodation during the harvest, so the farm would have been a very busy place.
(by Tom S.)

Lincoln Longwool sheep by Toby P.

We know about the wealth of the Newton farm from Isaac, the farmer's, will. It was written out for him by others and he simply put a cross at the foot as he could not sign his name. For Isaac, the time to pass on his wealth came all too soon as he died six months after his wedding to Hannah. Heavily pregnant, she became the Widow Newton, with the challenge of running a busy farm.

Her baby came in the middle of winter. Isaac Newton was born on Christmas Day 1642, very early in the morning. It was three months after his father had died so Isaac was posthumous. He was also premature, meaning he was born early. His mother said that he was so small he could fit inside a quart pot! Isaac was not expected to live as ate basic food. Being so small, he must have found the Manor House very busy and noisy as life on the farm went on. It would have been quite cold as well. When his mother was spending time with him, she would probably not have played any games with him. *(by Peter S.)*

It may have been hard, but at least it was secure. If the Royalists had come south of Newark, the farm at Woolsthorpe would have been a prime target for raids by hungry troops travelling on the Great North Road. When Isaac was two, Parliament's New Model Army defeated the Royalists at Naseby and in the following year, King Charles gave up Oxford and surrendered to the Scots. The Civil War was effectively over and, for a short time, Isaac would enjoy peace at home too.

When Isaac was only three years old, his widowed mother had an offer of marriage from a vicar called Barnabas Smith. She did not know what to do, so she consulted her brother who was also a local vicar, William Ayscough. She had tragically lost her husband and, although the farm was doing well, she had no other source of income. The Rev Barnabas Smith was wealthy so he could support Isaac's mum. There was only one problem, it was that Isaac couldn't stay with them as the vicar did not want him in his house in North Witham. So young Isaac had to stay at Woolsthorpe Manor with his grandma. This was very selfish on Isaac's mum's behalf. *(by Charlie W.)*

Every Sunday, the Grandparents would have taken young Isaac to the service at Colsterworth Parish Church just a short walk from the Manor.

Colsterworth Parish Church by Charlie L.

The sundial which Isaac Newton is said to have carved when he was nine years old now displayed in Colsterworth Church.

This is the village where Isaac Newton, aged nine, scratched a model sundial displayed on the stone wall of the church. The graffiti is now a popular visitor attraction in the village, on show in the church. Isaac was interested in sundials and in telling the time by them from an early age for the rest of his life. Surely his mother would have seen that something about Isaac was different, and that he wasn't cut out for farm work. *(by Daniel B.)*

Isaac was without his mother for the next eight years. When Barnabus Smith died, Hannah and the three new children she'd had with Barnabus came home to Woolsthorpe. The children were called Benjamin, Mary and Hannah. This must have been very difficult for Isaac as he now had to share his mother with his step-brother and step-sisters. *(by Jack S.)*

Much later, when he was about nineteen, Isaac wrote about this period of his life in the form of 'confessions'. He admitted to himself that his resentment and anger had poured out: *'threatening my father and mother Smith to burne them and the house over them'*. It would not have been difficult. North Witham where Smith held the living of the church was only a short walk away to the south of Woolsthorpe and thatched roofs were easy to light even by a child, especially one who was already interested in flying burning kites over the fields around the Manor.

While living at Woolsthorpe with his Ayscough grandparents, Isaac would have been of an age to learn to read and write. Boys and even some girls from rich families would have been taught by a tutor at home and it may be that the Ayscoughs arranged this for Isaac although there is a tradition that he was sent to a small school in the village or in one of the neighbouring villages.

He could have gone to a 'dame school' in Colsterworth or Skillington for his early years of schooling to be taught to read, write and to do simple maths by the older ladies of the village. The second explanation is that he was taught by his mother when she returned home when he was about eleven. 'Dame schools' were common around the Victorian era but most women didn't know how to read or write in the 17th century. *(by Sam L.)*

If it was his mother who taught him, he would have learnt at a significantly older age than other boys. Before they could be admitted to a grammar school in the Tudor and Stuart period, boys had to be able to read and write, and most boys went at around the age of seven or eight. At this age, Isaac was still at home with the Ayscoughs.

When his mother returned, she brought her late husband's considerable library, two or three hundred books, mainly works of theology and Church history but also a nearly empty notebook. This Isaac later used as the 'Waste Book' in which he recorded some of his earliest work on mathematics.

Two years after his mother returned, she was persuaded by his uncle to let him go and be educated at The King's School, Grantham. He was twelve when he went to the school. There was no entry exam required at that time, provided you could read and write, and the best thing was it was free if you lived in the town. But as Isaac lived at Woolsthorpe, his mother had to pay a fee to the Headmaster. *(by Tom S.)*

The King's School, Grantham, Today

The King's School from the Church Yard of St Wulfram's

The Old School 'in the shadow of the Parish Church spire'

The Master's House and apple tree at The King's School

Newton memorabilia held by Grantham Museum

5. Off to School

Grantham Grammar School existed long before it was re-endowed by Ropsley-born Richard Fox, Bishop of Winchester, in 1528. It became The King's School under a later charter of Edward VI. The Old School and the Master's House still stand in the shadow of St Wulfram's Parish Church spire in the heart of Grantham. The old buildings are still in daily use. In the 1640s no more than sixty boys from the town and surrounding villages would have come to the school to learn Latin and perhaps a little Greek until they left at about fifteen. Those deemed suitable then went on to University, most commonly from Grantham to one of the Cambridge Colleges.

As Woolsthorpe was seven miles from Grantham, Isaac's mother decided that he should live with a family in the town. She already knew the wife of the apothecary whose shop was on the High Street. Mrs Clarke shared Hannah Newton's experience as she had lost her first husband and had a posthumous son. So Isaac was sent to live with the Clarkes and their children in a house next to the George Inn. This is now the site of the ASK bistro next to the George shopping centre, a short walk from the school.

Lodging with the Clarkes provided Isaac with access to books and to Mr Clarke's work as an apothecary. He also got to know his step-children especially his step-daughter, Catherine. It seems that Isaac got on much better with girls than with boys at this stage, although the friendship with Catherine was one of the very few close relationships he had with anyone!

Life at The King's School would have been tough, especially for a lonely boy who was much older than the other new pupils and found it difficult to get on with the others. He must have stood out like a sore thumb, an obvious target for teasing.

Whilst living in Mr. Clarke's apothecary shop, Isaac became acquainted with Mr. Clarke's stepson, Arthur Storer, who also attended The King's School in Grantham. In 1655, Isaac was at the bottom of the class for everything. His school reports were also rather bad, describing him as inattentive and idle. His teachers said that he lacked motivation. Arthur Storer was said to be a bully, although he was in fact younger than Isaac. He may have called Isaac names such as 'newt-face'. One day, Isaac had had enough of Arthur, and everyone else, and got into a fight with him in the church yard behind the school. Young Isaac dragged Arthur by his ear and when he got to the church wall, rubbed Arthur's nose on the rough surface of the wall. From then on, Isaac's attitude changed. He was determined to be top at everything at school.
(by Jonathan B. and David T.)

Apart from these reports, there are no other records of Isaac's first five years at the school. In common with schoolboys in every age, he clearly wanted to leave his mark on the desks he sat in for hour after hour of learning Latin. On the window-sill in the schoolroom, now the Library, the name 'I Newton' is carved alongside many others.

Isaac was now immersed in the world of books. Not only did he have his school studies to keep him busy, but the Clarke household had books and there was a collection of books in St Wulfram's Church next to the school. These had been given to the town at the end of the C16th by Francis Trigge and are still housed in the small priest's room above the south porch. Mostly chained, and covering a wide range of subjects, they were nevertheless available to the people of the Town. Newton's Headmaster, Mr Henry Stokes, was one of the Trustees of this little library. He may well have encouraged Isaac to find his way to the top of the stone stairs to read them in the peace of the Parish Church and to escape from the taunts of his fellow pupils.

Amongst the books in Mr Clarke's collection was the 'Mysteries of Nature and Art' by John Bate, a collection of mechanical toys and models. Isaac was clearly fascinated by these and spent what little money his mother gave him to buy tools to enable him to make working models. One was a windmill which worked in the breeze and could even be powered by a captive mouse. He made a water clock which could be used to tell the time and folding lanterns for dark mornings. Some were light enough to be attached to the tail of a kite to float above the dark town, to the consternation of his neighbours. He even bought a small notebook in which to record details from Bate's book.

He also learned some of the apothecary's arts, the making and mixing of powders for remedies and ointments, the causes of common ailments and 'things hurtful for the eyes'. He began a more systematic approach to knowledge, recording in his notebook nouns beginning with each letter of the alphabet, 2,400 of them: *apothecary, armourer, astrologer, astronomer*………

On the night that Oliver Cromwell died in 1658, a great gale blew through Grantham. Isaac devised a way of measuring the strength of the storm by jumping first with the wind and then against it. By comparing the length of these jumps with what he achieved on a calm day, he could claim that the wind was *'a foot stronger than any he had known before'*.

Not only was he skilled in making and measuring things, his drawing skills developed too, thanks to plenty of practice on the walls of the Clarke's house. His room in the attic became a gallery of charcoal images of birds and plants, ships and people. There were portraits of the late King, Charles I, executed when Newton was six, of the poet, John Donne, and Mr Stokes. And everywhere, he wrote his name, and drew sundials!

6. Back to the Farm

After five years at school, Isaac's mother decided to call Isaac home from school to help to run the farm. After all it was his inheritance and role as 'Lord of the Manor'. Hannah Newton had struggled on her own for seven years and it was time for him to pull his weight. So he left his studies and returned to Woolsthorpe Manor. His mother did her best to train him in farming, assigning a trusted servant to instruct him, but Isaac resisted at every step of the way.

Isaac loved his schooling so to get back at his mother for taking it away from him, he left the sheep and cattle unattended to stray into other farmers' crops whilst he read a book or studied. One story tells us that Newton led a horse into town once and came back reading a book, carrying a bridle but without the horse; this shows how desperate he was to get back to his school life.

(by Marcus dB.)

He was taken before the Manor Court and fined for *'suffering his sheep to break ye stubbs'* and *'for suffering his swine to trespass in ye corn fields'*. His mother had to pay.

Letting the sheep stray, failing to bring the horse back in its bridle and being chased by the dog of an angry neighbour, Isaac fails as a farmer. By Charlie H-T, Oliver M. and Oliver W.

For nine months, Isaac behaved like the sulking teenager that he was. He was rude to his mother and his sister, fell out with the servants and refused to do what he was asked to do on the farm. He worked hard at demonstrating that he had not the slightest interest in becoming a successful farmer, or spending his life breeding sheep and sharing the company of the illiterate people of the local villages.

His mother consulted her brother, William Ayscough, as she always did at times of crisis. He talked to Mr Stokes at The King's School and they came up with a proposal. Isaac would return to school for a time to prepare for entry to University. Stokes was so keen to have his star pupil back that he agreed to rescind the fees his mother would have had to pay and even to have Isaac board in the Master's house, his own lodgings next to the school. Isaac was seventeen, much older than the other boys, but he must have been keen to escape from the farm. Given his behaviour for the previous nine months, his mother had to accept that what her servants said of Isaac was all too true. He was 'fit for nothing but the 'Varsity', and in the autumn of 1660, Isaac Newton went back to school.

This time, his experience was wholly positive. Mr Stokes made sure that he was well prepared for future study by encouraging him to learn some Greek. There is a notebook in the Lincoln archives which suggest that he may have taught Isaac some mathematics, but it is of a basic kind, the sort of practical calculations which would be useful to farmers working out areas of fields and yields of corn.

In the summer of 1661, he finally left The King's School, Grantham, and travelled south to Cambridge.

7. C17th Knowledge

For us, 'science' is such an obvious part of the modern world that it is difficult to understand the state of knowledge at the time of Isaac Newton's birth. In that year, the Italian astronomer and mathematician, Galileo Galilei died, a blind old man silenced by the Roman Catholic Church. He had developed a telescope powerful enough to see the moons of Jupiter and from his observations of their movements in orbit around the planet, he wrote about the motion of the planets.

Ever since the time of Aristotle in about 330 BC, it was accepted that the Earth was the fixed point at the centre of the universe and that the Sun and all the planets moved around the earth. The fact that the planets did not move in perfect circles was possible to explain by assuming that they moved in a series of loops. From the Earth, these made the planets seem sometimes to move backwards! The simple alternative, that the Earth was not the centre of the universe but was actually moving with all the other planets around the sun, had been proposed by later Greek astronomers. But because the whole structure of thought developed by Aristotle was accepted and taught, especially by the Church, his Earth-centred universe became the accepted truth. To challenge this was to challenge the authority of the Church and that is what Galileo did.

He knew from the work of the Polish monk Copernicus, published in 1543, that it was much easier to explain these apparently irregular movements if one assumed that the Sun was at the centre of the universe. Twenty years later, the Danish astronomer, Tycho Brahe, observed the movement of the Sun and planets with a new level of accuracy and found that the simple circular orbit model for planetary motion which Copernicus had suggested did not fit the best observations.

When, at the end of the C16th, the German astronomer Johannes Kepler was invited to work with Brahe, he found a way of resolving this conflict by proposing that the orbits might be ellipses rather than circles. The model worked and Kepler deduced three fundamental laws of planetary motion from Brahe's measurements. Firstly, that each planet travels in an elliptical orbit with the Sun at one focus of the ellipse. Secondly, that a line drawn from the planet to the Sun sweeps across equal areas of space in equal time, and thirdly, that for any two planets, the squares of their periods (the time for that planet to move once around the Sun) are proportional to the cubes of the distances from the Sun. Brahe's information fitted these laws, but they did not answer the question '*why?*'.

On four successive nights in June 1610, Galileo recorded in his notebook his observations of the moons of Jupiter.

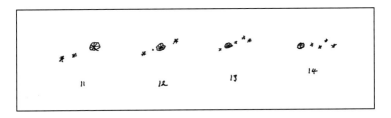

One night, he saw two spots of light on one side of the planet and none on the other. On the next night, there were three, one to the right, and on the next night, four, with three on the right. On the fourth night, Galileo saw that there were four spots of light, all to the right. He realised that the only way of explaining these observations was to suggest that the spots were reflections of light from four moons of the planet, sometimes visible from earth and sometimes in front of or behind the planet. If this was true for Jupiter and for the Earth's own moon, might it not also be true for the Sun itself, with the planets moving in orbit around the Sun.

The all-powerful Church forced Galileo publicly to deny his own theory. He had to agree that the Earth was as the Church taught, the static centre of the universe, but it is said that even at the time of this denial, he muttered under his breath *'and yet it moves'*.

There is no evidence that Newton knew anything about the work of any of these astronomers before going to Cambridge. His knowledge of mathematics was probably limited to the basic arithmetic and geometry which Stokes may have taught him and other sons of farmers in Grantham. Even the term 'scientist' would have been unknown to them as any who showed an interest in the material world and the heavens would have been called 'natural philosophers'- literally, lovers of knowledge about the natural world.

In the C16th and the first half of the C17th, the New World of the Americas was opening up to the European nations. The Spanish, Portuguese and Dutch had already established settlements in the Caribbean and in South America. The Portuguese and Dutch had trading missions on the coasts of Africa and India and in the Spice Islands of the Far East. The English were late starters but established their first permanent settlement in North America at Jamestown in 1607 and went on to settle 'New England' to the north in 1620 to create an English speaking North America. It was the great age of the 'joint stock company' with the Muscovy, Virginia and East India Companies raising money in the City of London to finance trading ventures in far distant parts of the known world.

Just as Europeans were establishing new settlements in exotic lands and astronomers were searching the heavens for new truth, so those with an interest in very small things were searching for explanations. William Harvey, doctor to James I and Charles I, discovered that all mammals reproduce by an egg being fertilized by a sperm, '*ex ovo, omnia*', although he is more famous for discovering that the heart pumped blood around the human body in a circulation system. The first microscopes were developed in Holland and in England and used to examine such novelties as the eye of a fly and the cellular structure of cork bark.

In London in the 1640s, a group of 'natural philosophers' began to meet together to discuss these exciting developments. At first, they called themselves 'The Invisible College', just a group of friends meeting in Gresham College in the City of London. They took up the ideas of the Elizabethan philosopher Francis Bacon, who has been called the father of scientific enquiry. Bacon entered Trinity College, Cambridge, at the tender age of twelve and followed the traditional course based on the work of Aristotle that had changed little since medieval times. After a distinguished career in Parliament and the Law which culminated in him becoming Attorney General to James I, Bacon fell into debt and out of favour, dying of pneumonia in 1626. It is said that he was experimenting on the preservation of meat by packing a chicken in snow. The chicken and Bacon got very cold but sadly, Bacon caught a chill and died.

He had published his *'Novum Organum'*, or New Instrument, in 1620 as a direct challenge to Aristotle's *'Organon'*. Bacon argued that we should develop knowledge by observation and experiment. *"Men have sought to make a world from their own conception and to draw from their own minds all the material which they employed, but if, instead of doing so, they had consulted experience and observation, they would have the facts and not opinions to reason about, and might have ultimately arrived at the knowledge of the laws which govern the material world."*

This became the basis of the scientific method and it was such observation and measurement that the members of The Invisible College met to consider. At the time when Isaac Newton was returning to King's School, Grantham in late 1660, a group of twelve men met to hear and discuss a lecture by Christopher Wren, then the Professor of Astronomy at Gresham College. Wren was typical of the 'natural philoshers' of his day, in that he was interested in all aspects of knowledge. We know him as one of England's greatest architects, but he was equally at home in Astronomy and Mathematics. John Wilkins was another member of the group, a theologian and mathematician with the distinction of having been first Warden of Wadham College, Oxford and later Master of Trinity College, Cambridge.

Robert Boyle was also a member. He was born in Ireland and educated at Eton College and travelled on the Continent rather than going to University. He set up a laboratory at home for experiments in early chemistry and then moved to Oxford where he continued the work which was eventually published as his '*Sceptical Chymist'*. Yet as any student of elementary physics knows, Boyle's name is always linked to the Law which shows that for a given mass of a gas, its volume is inversely proportional to its pressure.

Sir Robert Moray was the best connected of the twelve. A Scottish soldier, diplomat and freemason, he had served Charles I and fought for the Royalist cause. After the restoration of Charles II, it was Moray who told the King of

The Invisible College and persuaded him to grant the Royal Charter that established the group as the Royal Society in 1661. He became its first President. Moray was able to bring together ex-Royalists and ex-Parliamentarians to form a fellowship of the most distinguished intellectuals of the day.

From that start in 1660, the Royal Society has grown to a world-wide fellowship of 1400 outstanding individuals. They represent all areas of science, engineering and medicine and form a global scientific network of the highest calibre. To be elected to the Fellowship of the Royal Society, and so have FRS after your name, is still the highest accolade of the scientific community.

The Royal Society held a weekly meeting of natural philosophers and they appointed a Curator of Experiments to ensure that practical demonstrations and experiments took place at their meetings. The first to hold this post was Robert Hooke. *(by Tom H. and Ben M.)*

8. Studying at Trinity

In the C17th, England had just two universities, Oxford and Cambridge. Both consisted of semi-autonomous colleges which chose whom to admit and then to present to the University for 'matriculation' – admission to a course of study. Oxford had been the King's headquarters for much of the Civil War and Cambridge was at the heart of the Eastern Association, the power base for Parliament. Oliver Cromwell himself had been a student at Sidney Sussex College and was the Member of Parliament for Cambridge.

In the previous century, Cambridge University had been the home of the English Protestant Reformation with Cranmer, Latimer and Ridley all studying there or holding

College Fellowships before going on to become Bishops in the Anglican Church. When Queen Mary, better known as 'Bloody Mary', wanted to force England back to Roman Catholicism, she had all three condemned for heresy with many other Protestant leaders. She arranged for Cranmer, Latimer and Ridley to be burnt at the stake in Oxford because she knew that there would be less sympathy there. So the Oxford Martyrs were actually all Cambridge men.

Isaac was admitted as a sub-sizar to Trinity College, Cambridge University, on 5th June 1661. The term sizar basically means Isaac was little more than a servant. His jobs included serving the more wealthy students and Fellows. He cleaned, and possibly worst of all emptied chamber pots! The reason why people started as sub-sizars was because they were too poor to be admitted as students paying high fees. This is strange as Newton's mother could have afforded the fees – maybe she wanted to teach him a lesson by making him work as a servant.

(by Charlie W.)

He may even have been a servant to Humphrey Babington, given the Grantham connection. If this was the case, it was not too heavy a duty as Babington only spent about five weeks a year living at Trinity, attending for the remaining weeks to his duties as Rector of Boothby Pagnell, not far from Woolsthorpe.

Isaac Newton's name is recorded in the University Matriculation Book for July 8th 1661 when he took the oath to '*preserve the privileges of the University as much as in him lay, that he would save harmless its state, honour and dignity as long as he lived and that he would defend the same by his vote and counsel*'.

He bought what he needed, a lock for his desk, a pot of ink, candles and a chamber pot and moved into rooms at Trinity to take his place at the bottom of the College social scale. A young man who was used to having servants to look after him found himself required to attend to the needs of others, to serve their meals and only eat his own when the Master, Fellows and Pensioners had left the Hall.

It was hardly surprising that Isaac found it very difficult at first at Trinity. Just as at school, he was, at eighteen, much older than the other new students. Not least amongst the challenges was the requirement to share his room with another student who probably found him just as difficult to live with. He seems to have survived in this unhappy state for eighteen months before he met another student, John Wickens, who was having similar problems. They persuaded the College authorities to allow them to share rooms together, a friendship which was to last for twenty years. It was one of the very few close relationships that Isaac had and it served to sustain him throughout his time at the College.

His official studies would have been very similar to the course which Bacon had followed in Oxford a hundred years before. Aristotle was still the authority on logic, ethics and rhetoric and Isaac made a start on the '*Organon*' but he never finished it. As Aristotle had argued that everything that is in motion must be moved by something else, so there must be a 'first mover' – to the Christian theologians, this must be God Himself. Since the main function of both Oxford and Cambridge was to produce clergy and all College Fellows had to take 'Holy Orders', becoming priests of the Church, this curriculum served them well.

Isaac left a record of his own struggles with faith at this time. Writing in a form of short-hand which he devised for himself, he recorded his list of sins, including his earlier attempt to set fire to his mother's and step-father's house. He confessed to '*not living according to my belief*' although it is not clear just what he did believe at that stage. He admitted to himself that he had '*neglected to pray*' and that he had been negligent at the Chapel. He confessed to '*making pies on Sunday night*' and to '*squirting water on Thy day.*' Perhaps more serious from a religious point of view were his '*unclean thoughts, words, actions and dreams*' and on setting his heart '*on money, learning, pleasure more than Thee.*' Such Puritanical introspection would have set him apart from the pleasure seeking lives of his fellow-students.

With the small allowance provided by his mother, he had enough to meet his basic needs, with a little to spare on small luxuries, all recorded in meticulous detail in his notes. He began to lend small amounts of cash to other students whose drinking and playing at cards had run through their allowances from home. The accounts of his lendings included the names and amounts and when the debts were repaid. Such usury was frowned upon by the College authorities and cannot have made him popular with his peers, but at least he was useful to them, if isolated in every other way.

By the end of his first year, Isaac had filled pages of his notebook with his studies of Aristotle but there was little sign of a knowledge of much else. He was clearly bored by the requirement to study the works of the classical philosphers and at some point he noted down a Latin sentence which meant: '*Good friends are Aristotle and Plato, but a greater friend is truth.*' But where would he find this 'truth'?

Trinity College, Cambridge

Trinity Great Gate and the Chapel from a C17th College print.

Sundial and Great Court, Trinity College, Cambridge

Staircase of Newton's rooms in Great Court, Trinity College

The apple tree outside Newton's rooms in Trinity College.

For a junior sub-sizar, using the wide collection of books in Trinity College library would have been very limited as only Fellows had open access. But there were other sources of books. Each autumn, on the flat meadowland by the River Cam, traders from all over Europe came to sell their wares at Sturbridge Fair, said to be the largest open market in England, and the inspiration for John Bunyan's Vanity Fair in *'Pilgrim's Progress'*. Amongst all the other stalls, book sellers would offer whatever titles they had picked up on their travels, often on a weird mixture of subjects. When the good people of Grantham decided to invest £100 from Francis Trippe's legacy, they sent a man to Sturbridge Fair to buy books, any books!

So Isaac Newton went to Sturbridge Fair and used some of his allowance and the profits from lending money to buy books. He also bought a simple glass prism of no great quality, more of a child's toy than a scientific instrument. As he read his new books, he started to record his studies systematically, with notes entitled *'Quaestiones quaedam Philosophicae'* – 'some philosophical questions'. His reading widened to include the work of the French philosopher, Descartes, the writings of Galileo and the work of Robert Boyle. He read the philosophical works of Thomas Hobbes and Henry More, the 'Cambridge Platonist' who had attended The King's School, Grantham, in the generation before Newton.

But he was more and more interested in Mathematics, particularly geometry and trigonometry. He read Descartes's *Geometry* again, very carefully, to ensure that he had understood it. He found that he needed to know more of the classical geometrical proofs set out by the great Greek mathematician, Euclid. Descartes had set out a new way of relating geometry to mathematics by inventing a system of co-ordinates which are now called 'Cartesian'.

Any point in a diagram or on a map could be given an identity by measuring its distance from two staight lines at right angles to each other, called axes. Thus on a map, the co-ordinates of a point on the ground can be defined by the 'map reference', an 'easting' (which is the distance along the horizontal x axis) and a 'northing (which is the distance along the vertical y axis). For example, the reference for Woolsthorpe Manor is SK(the sheet of map for the area) 927245.

For simpler diagrams the two axes can be labelled 'x' (horizontal) and 'y' (vertical) and any point given a value of each. This system then allows for any line to be given a mathematical equation, such as the straight line through 0,0 which is defined as y = x for a line at 45°, or y-2x for a line with a steeper slope.

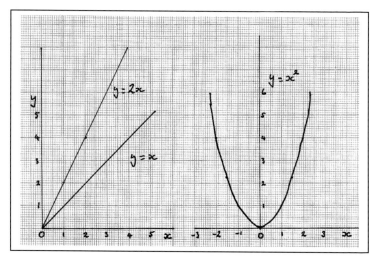

The system can then be applied to more complex lines such as curves and be used to describe how things move. The curve of a parabola is defined by the equation

$y = x^2$. A parabola turned upside down is the path which a ball follows when thrown forward up into the air, or the path a shell fired from a gun will follow. Using this system, it became possible to describe how things move in a new and precise way, by describing their movement in terms of a precise mathematical equation. What could be applied to throwing balls and firing shells could be applied to planets moving in the heavens. By adding a third axis, labelled z, at right angles to the other two, movement through three dimensional space could be precisely described.

From a very limited knowledge of mathematics, in his first year at Cambridge Isaac Newton had taught himself most if not all that there was to know in the field at that time. As a consequence, he had neglected much of the reading that formed the course he was supposed to be following. In this ill-prepared state, he had to face the first hurdle in his academic career at Trinity College, the examination for a scholarship. If he was ever to be elected as a Fellow, he must first be awarded one of the sixty four scholarships funded by the College. If he failed, he would have to leave after taking his degree.

The odds were against him. He had not shown any great distinction as a student in his first two years in the conventional studies required of him. As a sub-sizar, he had less chance of gaining influence with the Fellows, many of whom had come from Westminster School, whose 'old boys' year after year received about a third of the scholarships. As the elections were held at three yearly intervals, Newton had just one opportunity to compete, in 1664. Yet he had chosen to study Mathematics, a subject which had little status in the College. But he did have two allies. Humphrey Babington knew him from his past in Grantham and had encouraged him to come to Trinity in the first place. Babington was a influential senior Fellow and would become Bursar of the College, a role in which Newton helped him by devising tables to show when College leases should be renewed.

His other supporter may well have been the new Professor of Mathematics himself.

Rev Dr. Isaac Barrow, the first Lucasian Professor of Mathematics by Oliver M.

Before the Rev Isaac Barrow was appointed to the first Lucasian Chair, Cambridge had no professor of Mathematics. Barrow was already a Regius Professor of Greek, but Mathematics was a subject far more obscure than the language of Aristotle. Henry Lucas was the rich benefactor who had put up the money to establish the new post. Barrow had himself come to Trinity as a sizar and had worked his way through the levels of Scholar and Fellow to become the leading Cambridge mathematician of his day. When Newton's Tutor was looking for an examiner for his unconventional student, he sent Newton to Barrow, who was not at first very impressed.

Choosing to examine the young man on the proofs of Euclid, Barrow soon found that Isaac's knowledge of these was very limited. He would have been better equipped to explain Descarte's Geometry, but Barrow was convinced that this more advanced work could only be built on a thorough knowldege of Euclid's *Elements*. Nevertheless, he must have seen some spark of real ability in Newton's answers. When the list was published in the summer of 1664, Isaac Newton's name was included in the new scholars of Trinity College. He went again that autumn to Sturbridge Fair and bought a copy of Euclid for more systematic study, although he is said to have dismissed many of the proofs as *'so easy to understand that he wondered anybody would amuse themselves to write any demonstrations of them'*.

His scholarship gave him College security, an improvement in his status and easing of his budget. He could now dine in Hall with the Master, Fellows, Pensioners and Scholars, although it seems he frequently forgot to go. He received a small allowance for buying the right gown and another for his student expenses.

It was Isaac Barrow who suggested that Newton should widen his studies by reading Kepler's *Opticks*. It seems that he had already read Robert Boyle's *'Experiments and Consideration touching Colours'*. Alongside his thinking about mathematics in 1664 and 1665, Newton began a series of experiments on light and colour to test out theories for himself. Descartes had suggested that light was a form of pressure so that the light transmitted across the universe from the sun to the earth must be transmitted by pressure on matter, what Descartes believed was formed by whirling vortices which filled all space. If that was so, reasoned Newton, it should be possible to test the effect of pressure on the human organ which detected light. There was only one way to find out.

Firstly he poked sticks into his eyes which caused him to see several colours but gave him a lot of pain. Secondly, he stared at the sun for hours to see what effect it had on him. Because of this he had to stay in a darkened room for a week. *(by Jonathan B.)*

He was prepared to risk the sight of his own eyes to collect evidence. When he looked at the sun and then looked at a dark wall, he saw circles of colour which slowly faded – were they real or imagined? Light objects looked red and dark ones blue and he found that he could even imagine the colours without first looking at the sun.

His 'sticks in the eye' experiment was designed to increase the pressure on the surface of the eye by pressing on the stick. He noted that he saw *'several white, dark and coloured circles'* But these faded when the stick was held still. So where did the circles and the colours come from? In his notes on reading Descartes, he jotted down that perhaps Descartes was wrong. Light was not a form of pressure and the transmission of light from the sun did not need Descartes' vortices. There had to be a better explanation of what light really is.

Just before Christmas 1664, Newton was burning the midnight oil, or rather candle, in his rooms in Trinity, working on solutions to mathematical problems which he had set himself, when a terrifying sight was observed above Cambridge. Night after night, he stayed up to watch the comet that appeared in the darkened sky and then mysteriously disappear before the dawn. Comets were believed to be harbingers of disaster and news was already reaching England of a terrible pestilence spreading up from Italy and France. The plague was on its way north, just as the Black Death had come to England three hundred years before.

The 1664 comet seen by Newton above Cambridge, by Sam E.

In January 1665, Newton was due to take what would today be called his 'Finals', to be examined for his Batchelor of Arts degree. The University statutes required candidates to be questioned in a form of oral examination by Fellows, but by Newton's time, this had effectively been reduced to each College presenting candidates to the University Senate which had already agreed to the award of their degrees. Whatever really happened in Newton's case, it seems that his degree was not particularly distinguished. A scholar, who was at that time beginning to work on mathematical challenges far more advanced than anything attempted by other scholars across Europe, was awarded a second class and almost irrelevant degree.

9. Back to Woolsthorpe

In May 1665, Isaac's mother wrote to him from Woolsthorpe expressing her hope that he was well. Rumours were rife about the spread of disease and she must have been very worried. The Great Plague was not new to England. Ever since the 1360s, England had suffered outbreaks of the terrible disease almost annually. When in 1665 it struck London, the same horrors came back. It spread from London very quickly and by the summer of 1665, Cambridge was infected. As was common in this period, the rich fled and the poor were left to die. Isaac Newton fled home to Woolsthorpe Manor to get away from infection. But he carried on his studies and developed his ideas, taking with him his books and his small glass prism.
(by Jack S.)

Newton returned to Woolsthorpe as Lord of the Manor, Cambridge graduate and brilliant mathematician, but none of these seem to have impressed those living at the farm. We know little about his relationship with his mother at this time and nothing about his dealings with the farm workers as his sole focus seems to have been on his thinking. His mother remained at Woolsthorpe for another fourteen years until her death in 1679. She did provide small sums for him at Trinity College, but there are no more letters between them. When she caught a '*malignant fever*' from nursing Benjamin, Isaac's step-brother, Isaac nursed her and tried to ease her suffering by dressing her sores. But she died and was buried as '*Mrs Smith, widow*' in Colsterworth church yard, wrapped in a woollen shroud.

Isaac took over the chamber at the head of the stairs at the farm. Little remains of the furniture he used, apart from the small book press, the cupboard by the fire place, but he must have had an extensive library in the house.

Woolsthorpe Manor

The west front of Woolsthorpe Manor, birthplace of Isaac Newton

Newton's apple tree in blossom at Woolsthorpe Manor

Working in the orchard at Woolsthorpe Manor

Working in the Science Discovery Centre at Woolsthorpe Manor

It was still a busy farm-house, the ground-floor kitchen and the yard ouside a hive of activity. In the grounds in front of the house, an orchard provided apples for cider, fallens for the pigs and a quiet space for Isaac's reading and thinking. The link between Isaac Newton and the falling apple comes from the stories that he told to at least four other people, with slight variations on each occasion! What is clear is that a falling apple helped him to think about how the moon and the planets interact with each other in their movements in the heavens. An apple tree from Newton's time survived in the orchard for well over a century after his death.

In the garden of Woolsthorpe Manor you can still see the famous apple tree that is said to have borne the fruit that prompted Isaac's theory of gravity. At first Isaac could not work out the strength of this force and he didn't publish his ideas for over twenty years. The famous apple tree is now a shadow of its former glory. In the $C19^{th}$ it fell down and was left for dead after a terrible storm. But it sprouted again and cuttings were taken to grow new trees planted in Trinity College garden in Cambridge and in The King's School garden in Grantham. *(by Charlie W)*

Newton's relative, John Conduitt, told the apple story as Newton had told it to him.

While he was musing in a garden it came into his thought that the power of gravity (which brought an apple from the tree to the ground) was not limited to a certain distance from the earth but that this power must extend much further than was usually thought. Why not as high as the moon said he to himself and if so that must influence her motion.

The story of Isaac Newton and his apple continues to capture the public's imagination. In May 2010, the last NASA Space Shuttle mission included British astronaut, Dr Piers Sellers, on its twelve day journey into space and back. He took with him a small piece of the tree at Woolsthorpe from the Royal Society archives to celebrate their 350th anniversary and to acknowledge the debt still owed to Newton's work on gravitation.

In the winter of 1664, Newton had been working on mathematical problems in his room at Trinity. He knew that the system of co-ordinates could be used to give a mathematical desciption of a curve as an equation. He must have taken his notes of this to Woolsthorpe so that he could continue his studies even though he could not get back to Cambridge.

One place he is known to have visited is Boothby Pagnell, the village where Humphrey Babington was Rector. Here were more books which Newton could study and the peace in which to study them. In his own account, Newton refers to Boothby Pagnall as the place where he attempted to work out the area under the curve known as the hyperbola. This he did by working out the value of an infinite series of terms:

$$ax - \frac{x^2}{2} + \frac{x^3}{3a} - \frac{x^4}{4a^2} + \ldots$$

For this very complex calculation, he needed to use the idea of logarithms, invented by John Napier in 1614. Logarithms of numbers are the power of ten needed to make the number, and they make it possible to use addition and subtraction in the place of multiplication and division. For example,

$$100 = 10^2 \quad \text{so} \quad \log_{10} 100 = 2$$

and

$$\log(ab) = \log a + \log b$$

$$\log\left(\frac{a}{b}\right) = \log a - \log b$$

In his notebook, and using logs, Newton calculated the value of the area under a hyperbola given by the infinite series above, to fifty-five decimal places, rows of tiny numbers across the page. Having played with one infinite series, he worked out a general theorem or pattern for any infinite series, what we now know as the Binomial Expansion, the value of the sum of two numbers raised to any power n:

$$(a+b)^n = a^n + \frac{na^{n-1}b}{1} + \frac{n(n-1)a^{n-2}b^2}{1 \times 2} + \cdots$$

For the next three hundred years after Newton, 'logs' provided scientists and engineers with an aid to complex calculations. The slide-rule, used by all students of mathematics until recently, was simply a physical representation of 'log tables' in which two lengths could be quickly and visually added together. Today, few people use 'logs' or slide rules as the electronic calculator has made even the most complex calculation just a matter of pressing the right buttons, but the young mathematician in his isolated Lincolnshire farm-house trying to express the movement of an apple, and by extension, of the moon and the planets, had to do the calculation using the mathematical tools of his day. When there was no tool available, Newton had to devise one, just as he had invented tools to make models in the attics of the Clarkes' house in Grantham.

Curves, areas under curves, infinite series, the next problem in Isaac's list was to do with what he called the *'crookedness'* of curves. At any point, this *'crookedness'* could be turned into a straight line which touches the curve, the tangent to the curve, at that point. Under the newly invented microscope, even a curve would seem to be made up of a number of infinitely small straight lines, each a tangent to the curve at a point infinitesimally close to the next.

But each tiny tangent gives the slope of the curve at that point. Newton thought of a way of studying the slopes of these tiny tangents which he called his method of *'fluxions'*. He was in the process of inventing a new mathematical tool, what we now call the calculus. His *'fluxions'* idea did not survive because a better way of thinking about very small gradients to curves was invented about ten years later by the German mathematician, Gottfried Leibnitz. He called his method that of 'differentials' and used the expression: dy/dx for the gradient of the curve at any point. For example, for the curve which is described by the equation: the differential is:

$$y = x^5 \qquad \frac{dy}{dx} = 5x^4$$

As neither Newton nor Liebnitz told others of their work while they were thinking it through, who discovered the calculus first was an issue which rumbled on. It produced very acrimonious debates between them and their followers for decades to come. But both must have made the break-through quite independently of the other.

For Isaac, the value of the gradient could be calculated by using classical geometry but it was a very laborious process. By using his method of 'fluxions', he found he could calculate the gradient of any curve at any point. But lines and curves only show where things have been and where they are going, like the trails left in the sky by passing aircraft. Isaac realised that the process of movement itself was much more interesting, what he called *'resolving problems by motion'*.

If another graph is drawn in which the y axis remains distance but the x axis become one of time, a way of describing movement mathematically becomes clear. The slope or gradient of the line is the speed of the movement, constant if the line is straight and accelerating or decelerating if the line is curved. If his method of 'fluxions'/ 'differentials' was applied to these curves, a precise way of describing the rate of change of the speed of an object could be developed. He was getting close to turning the measurements so carefully made by Kepler of the movements of the planets, and by Galileo of the way objects accelerated as they fell, into precise mathematical expressions or 'laws'.

Between his 22nd birthday in 1664 and his 24th in 1666, much of which time he spent at Woolsthorpe because of the Plague, Isaac made his most significant mathematical discoveries. It was, as he described it much later in his life, the *'prime of my age for invention and I minded mathematics and philosophy more than at any time time since'*.

But there was still the puzzzle of 'why?'. He could describe the movements of the moon and planets and he *'began to think gravity extending to the orb of the moon and computed the forces requisite to keep the moon in her orb with the force of gravity at the surface of the earth ...... and found them answer pretty nearly'*.

If the apple fell to the ground because of gravity, why did the moon remain in the sky – was it something to do with the way it moved around the earth in an orbit? If the planets also remained in the sky, how did they influence each other, as they moved in well defined paths around the sun? Could the idea of gravity be much more than something that stops jumping boys from flying off into the air, but something that applies to everything, a universal principle?

For the moment, he had to be satisfied by *'pretty nearly'*, and his ever moving mind switched to other things.

Newton and the falling appple – artist's impression by Charlie H.

He knew from what he had read of the work of Boyle and others that light passed through a prism produced a range of colours but he had not seen this for himself.

He made a small hole in the wooden window shutters in his bedroom at Woolsthorpe so that a beam of white light hit a glass prism. It passed through the prism and shone onto a wall. However the patch of light on the wall wasn't white. In fact it was multi-coloured. But why?
(by Jonathan B.)

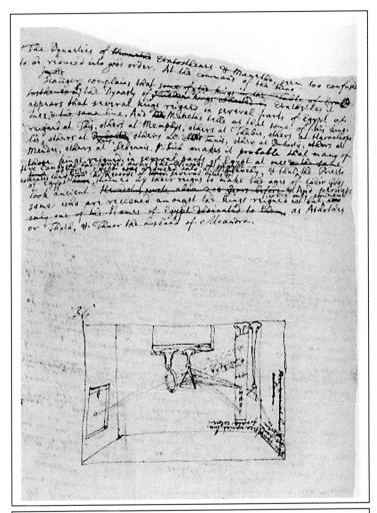

This a copy of a page from Newton's notebook showing the prism experiment at the bottom and notes on kings of Ancient Egypt at the top – typical of his range of interests in his reading and research. He continued to study optics, mathematics, theology, ancient history, alchemy, the books of the Bible and the writings of the Church fathers and many other mysteries throughout his life while working in a practical way on designing and making instruments such as telescopes.

White light, always thought of as 'pure light' seemed much more complex than had been thought. It appeared to be a mixture of different colours, separated out by the prism. But what happened when one of these colours was passed through a second prism? Would it be split up again? Newton did not have a second prism at Woolsthorpe and the authorities had cancelled Sturbridge Fair for 1665 on account of the plague. His light experiments would have to wait until he could get back to Cambridge.

10. Fellow and Professor

By March 1666, the terrible epidemic of the plague in London had died down and people were returning to the city. In Cambridge, the University re-opened and Newton returned to Trinity to continue his studies. He decided by June of that year to go back to Woolsthorpe where he stayed for almost another year, continuing to work on his mathematical problems.

After the long dry summer of 1666, the city of London was getting back to the serious business of making money when disaster struck again, just as those who saw the comet had said it would.

The Great Fire of London started on the 2nd of September 1666 at the house of a very good baker called Thomas Farriner, who lived on Pudding Lane. Thomas closed up his shop for the rest of the day and went upstairs, but he failed to ensure the oven was fully out. A bit of time passed and a single little, tiny spark flew out of the oven and landed on some fire wood. Within an hour the house was in flames. The baker and his wife, daughter and servant escaped from a high window but his maid got caught in the flames and died. Very quickly the fire spread to the wooden *houses across the street. Frantically the people ran from the fires but within a short time, about 13,200 houses and 87 churches were consumed. Amazingly, only 6 deaths were recorded.* *(by Gareth L.)*

By the time the great diarist, Samuel Pepys, had got out of his night-shirt and set out to watch the flames, the whole city was burning. Those who could save their belongings on carts, boats and on their backs were fleeing from the destruction. A few days later, Pepys wrote:

... by water to Paul's wharfe, walked thence and saw all the town burned, and a miserable sight of Paul's church, with all the roofs fallen .... Paul's School also, Ludgate, Fleet Street, my father's house and the church and a good part of the Temple the like......

Pepys had been admitted as a Fellow of the Royal Society the year before, by chance at a meeting on the nature of fire.
...was this day admitted, by signing a book and being taken by the hand by the President, my Lord Brunkard, ...a most acceptable thing to hear their discourse and see their experiments, which was this day upon the nature of fire and how it goes out in a place where the ayre is not free...above all Mr Boyle today was at the meeting and above him Mr Hooke, who is the most and promises the least, of any man in the world that ever I saw...,

Pepys, Boyle and Hooke, and Christopher Wren, met frequently at the Royal Society's weekly gatherings. They were held at Gresham College which had survived the Great Fire. Robert Hooke was given rooms in the College so that he could set up experiments for their meetings. He was the inventor of many practical devices including a universal joint, a telescope and a microscope.

His famous book *'Micrographia'*, a collection of exquisite drawings of objects seen through a microscope, was published by the Royal Society in 1665. After the Great Fire, Hooke and Wren collaborated on a plan to rebuild London. Although this was not adopted, Hooke did design and build the Monument to the Fire which still stands at the spot where it started and Wren's designs were used to rebuild many City churches including St Paul's Cathedral.

Throughout this time, while Newton was either at Trinity or at Woolsthorpe, no word of his activity reached London. He published nothing and apart from Isaac Barrow, no-one else seems to have been aware of his work. It was also time for him to compete at the next level at Trinity, the election to Fellowships.

There were nine places to be filled in October 1667 and some of the candidates already had an advantage as ex-Westminster scholars or were supported by letters from the King. Again, Isaac must have had Babington's and Barrow's support, for when the new Fellows were called to be sworn in by the Master, Newton's name was among them. It was a Minor Fellowship and to accept it, Newton had to swear that he would *'embrace the true religion of Christ with all my soul...and will either set Theology as the object of my studies and will take Holy Orders when the time prescribed by these statutes arrives, or I will resign from the College.'*

His fellowship once more gave him security at the College, rooms to use which he could rent out to others if he chose, and additional funds. He also received a fee for the tuition of a student, but there is no record that this student ever matriculated or graduated so Newton's duties must have been minimal. He did up the rooms he still shared with Wickens and when the time came for his BA to become a MA and for his Minor Fellowship to become a Major one, he could well afford the expense of buying cloth for another new gown.

Isaac Newton remained in residence at Trinity College for the next twenty-eight years. For much of that time, he was a recluse, seldom leaving his rooms, preferring to eat there alone rather than in Hall with the Master and Fellows. He had access to the College bowling green but never used it. Shortly after becoming a Major Fellow, he returned to his study of the nature of light, buying further prisms, lenses and boards into which he could drill small holes.

He repeated the experiment with the prism and tried shining the colourful rays of light through a lens. He found out that the once colourful rays of light turned back into normal plain white light. It was extraordinary and Isaac Newton was very happy. Then, Isaac had a wonderful thought, is there any way you could get a light ray and split it into the different colours of the rainbow then selecting one of those colours and seeing if it will split into the colours of the rainbow again.

Isaac needed to set up a second experiment in his room in Trinity College which at the time he considered as his laboratory. To do this he drilled a hole in the wall of his room then lined a prism up with the ray of light coming through the wall. This created the rainbow effect he had seen before. Then he used a board with a small hole to select one colour and lined another prism up with just the blue light. It shone onto the wall as pure blue light. Isaac had discovered you can only split light once.

(by Tom S.)

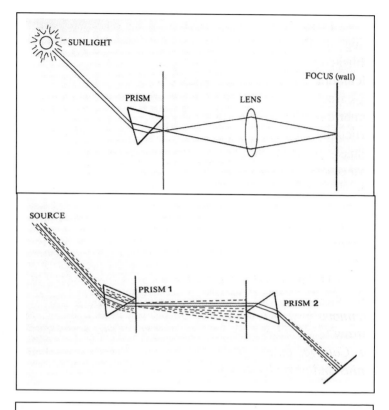

Newton's two experiments on the splitting of light with prisms

The story goes that Isaac Newton, working on laboratory light experiments, was repeatedly foiled by unwanted light from his cat pushing open the door. Not wanting to banish the cat, Newton cut a hole in the door, and attached a piece of felt to block out unwanted light. Since Newton and the cat were now happy, the world's first cat flap solved the light problem nicely. Later, when the cat had kittens, Newton cut several smaller sized holes in the door. However it didn't all go to plan as the kittens followed their mother through the large hole. *(by Andrew K.)*

In 1668, Nicholas Mercator, a French mathematician living in London and a Fellow of the Royal Scoiety, published a book *Logarithmotechnia*, which set out a way of calculating logarithms. The following year, a copy of this book was sent by John Collins in London to Isaac Barrow in Cambridge. As a Fellow of the Royal Society, Collins' particular interest was in mathematics and he made it his business to keep the leading scholars of his day aware of what others were doing. Barrow wrote back to Collins to say that a friend of his had a few days before showed him papers which set out methods of calculating logarithms '*like that of Mr Mercator concerning the hyperbola, but very general.....*' He promised to send him a copy.

The paper which Collins then received was in Latin and entitled *'De Analysi per aequationes numero terminorum infinitas' (On analysis by infinite series).* Barrow told Collins that this was by *'Mr Newton, a fellow of our College, and very young......but of an extraordinary genius and proficiency in these things'.*

Collins realised, from the dates of the writing of the book and the paper, that Newton had actually developed the method described about two years earlier than Mercator. Not only that but he had developed a much more general method well beyond what Mercator had achieved, but he, and other mathematicians, were catching up fast. When Barrow told Newton about Collins' letter and Mercator's book, Newton produced a more thorough account of his work to show that he had discovered the method first. He included his work on his method of '*fluxions*', and took it to Barrow for him to send on to Collins.

But then Newton, the *'extraordinary genius'*, got cold feet and pleaded with Barrow not to send the paper. He was afraid that his youth and the risk of his being shown to be wrong would work against him. But with Barrow's encouragement, he did eventually allow the paper to be sent to Collins on the strict understanding that he would read it and send it back. Collins knew, even with his limited understanding of mathematics, that the paper showed genius as Barrow had said. He did send it back but first copied it out without Newton's permission and sent it to leading mathematicians in England, Scotland, Holland, France and Italy.

Isaac Barrow had seen in Newton not just a great mathematician but also a solution to a personal dilemma. He was ambitious and genuinely interested in theology as well as mathematics. He had the opportunity to become a Chaplain to King Charles II, and also had his eye on the Mastership of Trinity College. But he could not hold these outside posts while continuing as Lucasian Professor of Mathematics. So he resigned his professorship and when the Trustees met to consider his successor, Barrow recommended the young Fellow of Trinity, Isaac Newton. In just over seven years, the sub-sizar with almost no knowledge of mathematics had become the Lucasian Professor at Cambridge University.

His new post carried one other advantage for Newton. On becoming a Fellow of Trinity, Newton had agreed that he would 'take Holy orders', become a priest, in due course. But Newton was becoming increasingly convinced that he could not, in all honesty, agree to the Church's doctrine of the Trinity. He read and re-read the Scriptures. He studied the books on the Early Church in his step-father's library. More and more, he convinced himself that God could not be Three Persons. He was moving towards the idea that had been put forward by an Eastern theologian, Arius, that God was not three but one.

The dispute had been settled at a Council of the Church in 325 AD, when the Western Roman Church declared that the doctrine of the Trinity was the truth, although much of the Christian Church in the East remained convinced that Arius was right. For a young Fellow of the Cambridge College set up in honour of the Holy Trinity and committed to a future in the Trinitarian Church of England, the risk of being seen as an Arian heretic was acute.

But for Newton, there was a way out or at least a way of putting off the inevitable. By accepting the professorship, he would have status and security for as long as he chose not to be ordained. This he could delay for up to seven years so it was only putting off the time of his ordination or resignation from his Fellowship, but it gave him a breathing space. He could make a start on his professorial duties which included giving a weekly lecture.

Newton chose to lecture on his work on Optics, one of the subjects specified in the Lucasian statutes. It seems that few students turned up to hear him and those who did seemed not to have gained much from the experience. Soon, the Lucasian professor was lecturing to an empty room. Not that that discouraged Newton, he went on with the lectures anyway.

His interest in optics had led him to to think about lenses and telescopes. He knew that the telescopes available to those who wished not just to see the stars and planets but to measure aspects of their size and movement had a major flaw. Just as the prism split light into colours, so the lens acts as a prism towards its edge, splitting light and producing a blurred image with coloured edges. This is what we now call chromatic aberration. However good the telescope was, this inherent feature of lenses would limit their usefulness.

Newton realised that a mirror would avoid this problem. Because no light passes through it, there would be no splitting into colours. A mirror could be made with a concave surface and another smaller mirror used to direct the rays of light from the mirror through 90^0 into an eyepiece. This is the principle of the reflecting telescope. So Newton made a flat cylindrical block of tin/copper alloy, and ground and polished its surface until he had made a concave mirror of exactly the right curvature. He then mounted it in a brass cylinder, again of his own making, fixed a small mirror at the focus of the mirror and supported the whole assembly on a stand. It was only about six inches long but this simple relecting telescope was, to his delight, more powerful than a refracting telescope over six feet long. He had made it by hand and had himself made the tools he needed.

He showed it to Isaac Barrow and word of the new telescope got down to London. Soon the Fellows of the Royal Society were asking to see it and in December 1671, Barrow took it down to show them. It caused a sensation and the Secretary of the Royal Society wrote to Newton concerned that the new invention should be 'protected' before it fell into the hands of foreigners. They proposed that a description should be sent to the great Dutch instrument maker Huygens to ensure that Newton got the credit. They elected Isaac to the Fellowship of the Royal Society in January 1672 on the strength of his telescope, not his mathematics. A London craftsman was commisioned to make two more such telescopes, one four and one six feet long, but making suitable mirrors proved to be beyond his skills.

Replica of Newton's 6 inch reflecting telescope from The King's School

In February of the same year, Newton sent an account of his work on light and colours to be read to the Society. Although well received by most of the Fellows, this letter provoked Robert Hooke, who considered himself the leading authority on colour, into harsh criticism. It was the beginning of a conflict between Hooke and Newton which would continue for the rest of their lives. Hooke claimed for example that Newton was not the first to make a reflecting telescope – he had himself worked on the principle of reflection and made one, although he had shown it to no-one.

The acrimonious correspondence on the theory of light rumbled on. Was light made up of 'particles' as Newton suggested or was it a matter of waves which Hooke favoured? Newton's letters to the Royal Society included the observation that Hooke's theory was *'not only insufficient, but in some respects unintelligible'*. We now know that light can be considered to consist either of particles or waves so in a sense they were both right. But, at the time, Robert Hooke had to endure the reading out of Newton's letters about him, full of hatred and rage, to the meetings of the Royal Society.

At one point, Hooke wrote to Newton with what seemed to be an attempt at conciliation. Newton replied in similar tones commending Hooke for his *'true philosophical spirit'* and adding: *'If I have seen further, it is by standing on the shoulders of giants.'*

This appears to be a compliment, but Robert Hooke had a deformed back and shoulders. It may actually have been a hidden insult referring to Hooke's deformed body. There was no love lost between them. After Hooke's death, Newton burnt all the paintings of Hooke he could get his hands on. *(by David T.)*

Newton eventually wrote to the Royal Society Secretary that he wished to withdraw as a Fellow as he could *'neither profit from the discussions nor take part in meetings, being in Cambridge'*. He was asked to reconsider and offered free membership but Newton simply ignored the letter.

In the following year, Newton did go to London as a representative of Trinity College at the installation of the King Charles II's illegitimate son, James, Duke of Monmouth, as Chancellor of Cambridge University. Not until 1675 did he get round to attending his first meeting of the Royal Society in Gresham College and at later meetings that year, he met Robert Boyle.

Robert Boyle by Oliver M.

11. Theology and Alchemy

Just before this visit to London, Newton had to face the date on which he had to take Holy Orders or resign his fellowship. There was one last option, a direct approach to the King. Dispensations from College and University statutes could be granted by the reigning monarch and again, it must have been Isaac Barrow, now a Chaplain to King Charles II and also Master of Trinity College, Cambridge, who came to Newton's aid. The King graciously agreed that the holder of the Lucasian Professorship of Mathematics should be exempt from the requirement to take Holy Orders *'to give all just encouragement to learned men who are and shall be elected to the said professorship'*. There was no mention of which 'learned man' this referred to. But while he remained Lucasian Professor, Newton could hold his Trinity Fellowship and remain in his base at Cambridge for as long as he liked, without taking Holy Orders and without any need to explain his real reasons.

Newton certainly wanted to stay at Trinity. He had moved to rooms in the College which linked to a wooden shed outside which he used as an alchemist's laboratory. John Wickens and he spent days there distilling liquids and dissolving metals. It had become his sanctuary and might well have been his tomb. Many of his experiments involved working with highly poisonous materials including compounds of antimony, lead and mercury. More than once, when he left his vessels bubbling away to visit the Chapel or to find a book, he returned to find his room on fire, his papers burnt, the meticulous records of his work lost for ever.

For about ten years, Newton remained at Trinity almost without outside contact. Not only was he immersed in his alchemical reading and experiments, he was also searching through the Scriptures and the works of the Early Fathers to find the truth. He studied prophecy in the books of Daniel and Revelation and wrote a history of the church to show how events had fulfilled biblical prophecies. He taught himself Hebrew to better understand the prophesies of Ezekiel and became obsessed with the detailed measurements of the Temple in Jerusalem.

These studies remained largely unknown to Newton scholars until the discovery of a collection of his papers which was bought by the Cambridge Economist John Maynard Keynes, just after the World War II. We now know that Newton was as much interested in alchemy and theology as he was in mathematics. But his name will ever be associated with the laws which govern the movement of planets rather than his fruitless search for the elusive 'Philosopher's Stone' which turns all things to gold.

During the increasingly acrimonious correspondence with Robert Hooke, the Secretary to the Royal Society,

Henry Oldenburg, had tried hard to placate Newton, but Oldenburg died in 1677 and the Society elected Robert Hooke as Secretary in his place. This effectively cut Newton off from the Society as his conflicts with Hooke had been so bitter. Hooke meanwhile published his work on spiral springs – what we now know as Hooke's Law, that the extension is proportional to the weight applied. He also wrote papers on his work on comets and gravity, both subjects close to Newton's heart. From then on, even after Hooke's resignation from the Secretary post in 1682, it was inevitable that they should continue to compete and quarrel.

Over the winter of 1680-81, another comet appeared over England. Newton made observations of its movement until it disappeared again in March. This was later called 'Halley's Comet' after the astronomer, Edmund Halley. He had set up an observatory on the South Atlantic Island of St Helena and had produced a wealth of accurate measurements of the movements of the planets – just what Newton needed to develop his theories of planetary motion.

Edmund Halley by Sam E.

Halley knew of the laws Kepler had derived from Brahe's measurements and was trying to work out why planets should move in elliptical orbits. They all knew that this was the case, but how could it be explained – the why? question again.

In 1684, some time after Newton had been made Lucasian Professor of Mathematics at Cambridge, Edmund Halley met with Robert Hooke and Christopher Wren at the Royal Society. Halley placed a bet with Newton's rival, Robert Hooke, that he couldn't explain why the planets and comets orbited the Sun in ellipses. As usual, Hooke said he could. Halley gave him a few weeks to reply, but when he couldn't write out the proof, Halley went up to Cambridge to Newton for some help, and Newton agreed. He explained that he already had the proof. However, he failed to find the papers. But Newton was so determined that he did the calculations again and wrote down nine pages of complicated mathematics. Newton sent the pages to Halley. Hooke never came forward, and Halley won the bet.

(by Peter S.)

Halley now knew that Newton had indeed developed a method of showing why the planets moved in elliptical orbits around the sun. The explanation was set out in this nine-page paper entitled 'De Motu Corporum, in Gyrum', or 'On the Motion of Bodies in an Orbit'. Halley realised that this was a huge breakthrough and went back to Cambridge to persuade Newton to present the paper to the Royal Society, which Newton agreed to do. Months later it did arrive in London and was received by the Royal Society, but by then Newton had already begun work on a much more general account of the mathematical principles behind the motion of the planets, a sytematic account of the movements of all bodies in the solar system.

He had shown years before that he could describe planetary motion in terms of relatively simple mathematical equations. The crude measurements of the planets' movements fitted the mathematics *'pretty nearly'*. But he knew that as soon as he published his work, it would be subject to the most exacting scrutiny and *'pretty nearly'* would not be good enough.

He needed better measurements, from the newly established Royal Observatory in Greenwich and from observers scattered across the new settlements in distant lands. He pestered the Astronomer Royal, John Flamsteed, for more and more accurate information on the movement of the moon and planets. Flamsteed was willing to help but must have been increasingly exasperated by Newton's insistent demands.

An irritated John Flamsteed by Oliver W.

He was in touch with astronomers right across Europe, in Coburg in Germany, Avignon and La Fleche in France, Rome and Padua in Italy. He received letters containing useful information from Jamaica, Brazil and Ballasore in the East Indies. Measurements were sent from Boston in New England and, best of all, from one Arthur Storer in the new English Colony in Maryland. This was the same young man whose nose Newton had rubbed on St Wulfram's Church and who had gone to Maryland with his sister. In the clear night sky of Maryland, he saw the same comet that Newton had seen in Cambridge. Not only did he see it, but he took accurate measurements of its movement which he sent back to Isaac.

As the information came back to Newton, he could see that *'pretty nearly'* was getting closer to *'precisely'*. The information, what we now call data, fitted the mathematics. He was confident enough to go on record and to face his critics.

12. The 'Principia'

From Halley's visit to Cambridge in August 1684 to the Spring of 1686, the sole focus for Newton's life was the text of the book for which he would become famous. He began to build an intellectual structure every bit as complex as the great cathedral which Wren was building at St Paul's. Starting with the nine mathematical pages of '*De Motu*', Newton needed first to define what he meant by words which described terms in his equations, words like mass, velocity, inertia and force. He set out eight definitions together with the relationships between them. He defined mass for the first time as the product of a body's density and volume, and weight as the force exerted on the mass of a body by the pull of gravity. His distinction between the two enabled him to use exact terms, but the confusion between them in everyday speech remains to this day. We talk about 'wanting to lose weight', when what we really mean is 'wanting to lose mass'. We could lose weight, but this would, strictly speaking, mean going to the moon where gravity is much less than on Earth. Eating fewer cream cakes is a much easier option! Newton's definition did not cover such everyday parallels, but he did add useful '*scholia*' or notes at the end of each section.

He then built a framework of Axioms or Laws, which have become 'Newton's Laws of Motion':

LAW I.
Every body perseveres in its state of rest, or of uniform motion in a right line, unless it is compelled to change that state by forces impressed thereon.

LAW II.
The alteration of motion is ever proportional to the motive force impressed; and is made in the direction of the right line in which that force is impressed.

LAW III.
To every action there is always opposed an equal reaction: or the mutual actions of two bodies upon each other are always equal, and directed to contrary parts.

Modern physics textbooks may use slightly different language, but the meaning is the same. In any case, all versions are translations as Newton wrote the whole text of his work in Latin, the commom language of European educated men. To his Laws, he added '*corollaries*' or things which could be inferred or deduced from the laws. Having constructed his foundations, Isaac Newton launched into Book I, a much expanded version of '*De Motu*'. He set out fourteen sections, each made up of a sequence of '*Lemma*' or proven statements, which he then used to prove another statement. From these, he derived propositions and problems getting more and more complex, but each building on the one before. Each is supported by geometrical diagrams and ever more advanced mathematics. The last of the sequence he entitled XIV: '*Of the motion of very small bodies when agitated by centripetal forces tending to the several parts of any very great body*', which is not far from a small apple falling towards a large Earth.

His sequence of proofs spilled over into a second volume, Book II, covering nine more '*Lemma*', each with propositions and problems on the movement of bodies. They included bodies moving through '*resistant fluids*' and ended on the circular motion of fluids themselves. The structure of the text is similar to that of Euclid's *Geometry*, each section building on the last until a mathematical framework is established to cover all aspects of moving systems. Having considered the movement of planets, he considered the case of comets and showed how helpful all those observations sent in by others, had been:

'Proposition XLI. Problem XXI. From three observations given to determine the orbit of a comet moving in a parabola:
Observations: Comet of year 1680 observations of Flamsteed, corrected by Dr Halley, some observations of my own, by Mr Pound, although my micrometer was none of the best, yet the errors in longitude and latitude...scarcely exceed one minute....
...The same day, Nov 19th, Mr Arthur Storer, at the River Patuxent near Hunting Creek in Maryland, in the confines of Virginia, in lat 38 ½ at 5 in the morning (that is at 10 hr in London) saw the comet above Spica and very nearly joined with it...... ......Moreover, Dr Halley observed that a remarkable comet had appeared four times at equal

intervals of 575 years, the month after Julius Caesar was killed, 531, 1106, 1680....'

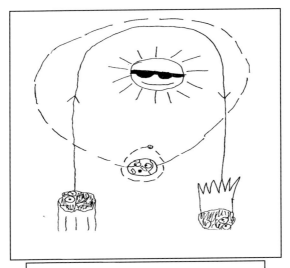

The movement of a comet by Charlie H-T.

For a time, Newton strayed from the mathematical purity of his argument in *Principia* into these personal anecdotes on comets. He showed that they behave not as mysterious harbingers of good or bad news but as bodies moving in the heavens with as clearly defined and predictable orbits as the planets. They come into view from the Earth thanks to their reflection of light as they pass close to the Sun and then move off again into space, only to reappear many years later.

Halley's Comet also has an elliptical orbit which causes it to disappear into space until it comes round again every 75-76 years. It was last seen from Earth in 1986 so it won't reappear again until mid 2061. (*by Sam L.)*

Newton completed Book II of *Principia* and then paused for breath, to remind the reader of his purpose:

'In the preceding Books I have laid down the ......mathematical ......principles... the laws and conditions of certain motions, and powers or forces; but, lest they should have appeared of themselves dry and barren, I have illustrated them here and there with some philosophical scholiums...'

He then started on Book III, *'the Systems of the World'*, with a warning. He had planned to write this *'in a popular method, that it might be read by many'*, but he decided to avoid disputes with those lesser mortals who *'had not sufficiently entered into the principles'* in Books I and II and so *'could not easily discern the strength of the consequences, nor lay aside the prejudices to which they had been many years accustomed'*.... So he chose to make Book III accessible to *'readers of good mathematical learning'*, but not too accessible! He suggests that his readers should first master the contents of the first two books.

For those who could not find the time to do this, he suggests *'...it is enough if one carefully read the Definitions, the Laws of Motion, and the first three Sections of the first Book. He may then pass on to this Book, and consult such of the remaining Propositions of the first two Books, as the references in this, and his occasions, shall require.'*

Much of Book III consists not of mathematical formulae but of tables of the information gathered from the sources he consulted, tables on the movement of Mercury, Venus, Earth, Mars and Saturn, against the framework of stars in the night sky. Newton referred back to Books I and II to show how the mathematical relationships he developed there apply to these movements and explain the Laws which Kepler developed. He explained the movement of the Moon in relation to the Earth and began to develop a universal principle of gravity. *'That there is a power of gravity tending to all bodies proportional to the several quantities of matter which they contain.'* He went on to show that this *'power of gravity'*, what we would now call a gravitational force, *'varies inversely with the square of the distance between the bodies'*.

In his final section, Newton dismissed Descartes idea of *'vortices'* and backed Boyle's notion that the planets move un-resisted through the vacuum of space. He set out a mathematical structure which covered the behaviour of the six known planets with their ten moons and the comets which come and go through space and decided that this whole system was no accident. As the believer that he was, Isaac Newton declared

'This most beautiful system of the Sun, planets, and comets, could only proceed from the counsel and dominion of an intelligent and powerful Being. And if the fixed stars are the centres of other like systems, these, being formed by the like wise counsel, must be all subject to the dominion of One... Lord over all...'

The writing of *Principia* was now finished but, before it could be published, the text in Newton's own small handwriting had to be copied out into legible form. John Wickens, Newton's first companion at Trinity had moved on but another student, Humphrey Newton had joined Isaac. They were not related although Humphrey had also been at The King's School in Grantham. As Newton finished each section of his work, Humphrey produced a fair copy word for word.

In late 1686, Isaac finished his first two volumes but heard that Robert Hooks had spread the word around that Isaac pinched other scientists' work. This upset him a lot and he refused to release his final volume. So, in the end, Halley brought Christopher Wren to Isaac who told him that Hooke was trying to make him feel bad. Isaac rewrote the third volume so that it was so complicated that even Robert Hooke could not follow it. Isaac finished Principia in 1687 after a year and a half of solid work.   *(by Jonathan B.)*

The reason Principia took Isaac Newton so long to publish was because he didn't like showing his work to other scientists and to the Royal Society. He could have made the book a long time before it was actually published, as he had figured out most of the things that the book would contain. He only agreed when Edmund Halley, who had recently made a complex map of the stars of the Southern Hemisphere, asked him to publish it. With Halley's help and encouragement along with the Royal Society's backing, the first books were published in 1687 but only five hundred copies were printed. He did this all when he was forty five years old.   *(by Sam L.)*

Edmund Halley had encouraged Newton throughout this period and as the text arrived in sections at the Royal Society, he persuaded the Fellows to publish the whole. But they had a financial problem. Although Hooke's *Micrographia* had sold well, their most recent venture, a book on *The History of Fishes*, had not. They simply did not have the money to publish Newton's work. Halley came to the rescue and offered to pay for the printing himself.

Reactions to the book were mixed, but less critical than Newton had feared. Halley helped again by writing an ode on '*this splendid ornament of our time and our nation,*' and a very positive review.

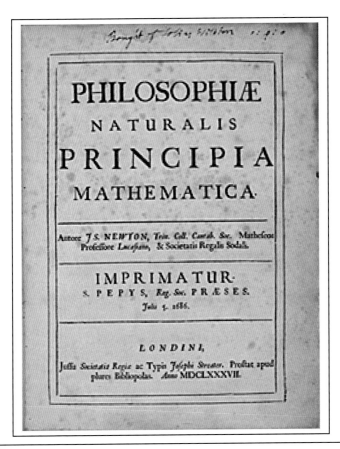

The front page of the first edition of Newton's Principia, published by the Royal Society under Samuel Pepys' Presidency in 1687

An unknown Cambridge student is said to have commented on seeing Newton *'there goes the man that writt a book that neither he nor anyone else understands'*. But he was wrong. The book's greatness was recognised even by those who had quarrelled with him. Leibnitz later commented that *'taking mathematics from the beginning of the world to the time of Sir Isaac, what he had done was much the better half'*.

13. To London and the Mint

Two years before the *Principia's* publication, King Charles II died and was succeeded by his brother, James II. He had married a Catholic princess and wanted to favour the appointment of Catholics to positions of influence. This brought him into direct conflict with centres of Protestantism, including the Universities. James tried to force Magdalen College, Oxford, to appoint a Catholic President and instructed Cambridge to award an MA to a Catholic priest. Cambridge University appointed Isaac to be a representative to petition the King. Newton found himself in some danger. To oppose the King was little short of rebellion and an actual rebellion led by the Duke of Monmouth had already been put down with great savagery. Newton and his colleagues were taken before the infamous Judge Jeffreys and were lucky to escape with an admonition to *'go home and sin no more'*. Newton had come out of his private world of Trinity College and had his first taste of public life.

When, in the following year, King James II and his wife had a son, England faced a future under a Catholic King. Leading Protestants wrote to the Dutch leader, William of Orange, whose wife had a claim to the English throne as the daughter of Charles I and sister to James II. They were invited to invade England and landed with an army at Torbay in Devon – the Glorious Revolution of 1688. James, fled and England was assured of a Protestant future.

When the Cambridge University Senate met in January 1689 to choose two representatives to send to London to take part in a Convention to set up a new Parliament, they chose Isaac Newton. He was rapidly becoming a nationally important figure. He had a very flattering portrait painted by Sir Godfrey Kneller, the leading artist of the day, and dined with William of Orange. When the Convention became the first Parliament under William and Mary, Newton sat as a Member for the University of Cambridge. There is no record of his contribution on the floor of the House of Commons – it is said that he did not speak at all in debates, although he did take part in the celebrations of the coronation of William and Mary.

If he was to stay in London, he needed a network of friends. He still had financial security from his Trinity Fellowship and Lucasian Professorship even though he was seldom there. In London, he could attend the Royal Society meetings and clearly enjoyed meeting the leading intellectuals of the day including the philosopher John Locke and the astronomer Christopher Huygens. He also met a young Swiss mathematician, Nicolas Fatio de Duillier, with whom Newton had a close friendship until they abruptly parted company in 1693. When this happened, Newton underwent some sort of mental breakdown, a *'discomposure of head or mind or both'* and wrote strange letters to both Locke and Pepys. Both did their best to help him and it now seems likely that Newton was suffering from a combination of over-work and mercury poisoning. With the end of the friendship with Fatio, Newton decided to make a fresh start and to seek a new role in London which had little to do with mathematics, alchemy or theology.

One of his Trinity College associates, Charles Montagu, became Chancellor of the Exchequer, the first to hold the post. He had in his gift the post of Warden of the Royal Mint, responsible for the production of the nation's coinage. In 1696, he offered it to Newton. The French writer, Voltaire, visiting London in the 1720s claimed that this was because Montagu had fallen in love with Newton's very attractive niece, Catherine Barton, who was acting as Newton's housekeeper at the time. But Catherine had not yet even met Montagu and the story seeems to be a product of Voltaire's over-active imagination.

The post of Warden of the Royal Mint was a surprising role for a leading scientist, but one in which Newton was as efficient as he was in all his other roles. In the late 17th century, the activities of 'clippers and coiners' were a very large problem. 'Clippers' would cut tiny bits of gold off the side of a coin and then use the bits to mix with copper to make fake coins, 'coining'. These they could sell for a good profit to be used as real coins. Isaac made it harder for the 'clippers' to cut coins by having them made with a milled edge. On some £2 coins today there are the words ' standing on the shoulders of giants' on the edge of the coin, a direct link with Newton's time at the Mint as this is the phrase created by Isaac, perhaps to annoy his enemy in the Royal Society, Robert Hooke! *(by Jack S.)*

Newton did not just re-organise the production of coins at the Royal Mint, he also went after the individual 'coiners', setting up a network of informers to gather evidence for their conviction. The most notorious was William Chaloner and, for a time, Chaloner managed to evade arrest by betraying other coiners and having some of Newton's informers murdered. Eventually, Newton had a cast-iron case against Chaloner, who was convicted at the Old Bailey of High Treason, the consequence of conterfeiting the King's coinage, and condemned to die. Chaloner wrote one last pathetic letter to Newton pleading for mercy, but he was taken to Tyburn on a sledge and hung, drawn and quartered, the death of a traitor.

Queen Anne came to the throne in 1702. Newton sat for another portrait by Kneller. In the following year, Robert

Hooke died. Newton was elected President of the Royal Society, a post he held until his death. In 1704, Newton published his book on Optics, most of which came from his Cambridge lecture notes of years before. Queen Anne came to Cambridge in 1705, and Newton became 'Sir Isaac'.

This was probably not for his mathematical or scientific work. Isaac was a supporter of the Whigs (who were Liberals). The other party, the Tories, had fallen out with the Queen and she decided to knight as many well-known Whigs as possible to get their support.

(by David T.)

Newton lived on in London, presiding over the Royal Society, publishing new editions of *Principia* and *Opticks*. His niece, Catherine Barton, married John Conduitt who collected stories about Newton and acted as a source for future biographers. Newton also met a doctor called William Stukeley, a Fellow of the Royal Society, who came from Lincolnshire.

William Stukeley by Oliver M.

We have to thank Conduitt and Stukeley for much of the detail of Newton's life. On March 2nd 1727, Newton attended his last meeting of the Royal Society, and his health collapsed. He had first suffered from kidney stones five years earlier but this time, the stone could not pass. Such was his pain that, in Stukeley's words *'the bed under him, and the very room, shook with his agony'*. He refused the Last Rights of the Church and died in the early morning of March 20th 1727. His body lay in state in Westminster Abbey before being buried in an elaborate tomb in the nave. England had lost her greatest scientist.

Newton's tomb in Westminster Abbey by Sam E.

He was eighty-four when he died. The French mathematician Joseph-Louis Lagrange said that Isaac Newton was the greatest genius ever and 'the most fortunate man for we cannot find more than once a mathematical system of the world'. *(by Tom S.)*

Sources

We found the following books and web-sites particularly useful in researching this story:

Ackroyd, Peter *'Newton'* Vintage 2006
Atkinson, Mary *'Genius or Madman? Sir Isaac Newton'* Scholastic 2008
Gleick, James *'Isaac Newton'* Harper Perennial 2003
Hawking, Stephen *'On the Shoulders of Giants: The Great Works of Physics and Astronomy'* Running Press 2002
Latham, Robert (Ed) *'The Shorter Pepys'* Penguin Books 1985
Newton, Sir Isaac and Motte, Andrew *'Principia'* Great Minds Series University of California Press 1995
Poskitt, Kjartan *'Isaac Newton and his Apple'* Scholastic Children's Books 1999
Westfall, Richard S *'Never at Rest, a Biography of Isaac Newton'* Cambridge University Press 1983
White, Michael *'Isaac Newton: The Last Sorcerer'* Fourth Estate 1997

www.isaacnewtonsoldschool.org/
www.kings.lincs.sch.uk/
www.nationaltrust.org.uk/main/w.../w-woolsthorpemanor/
www.newton.ac.uk/newtlife.html
www.newton.ac.uk/newton.html
www.newtonproject.sussex.ac.uk/
royalsociety.org/
350.royalsociety.org/
www.trin.cam.ac.uk/

and many others

Thanks

This book is a 'Books with Schools' Project within the series published by Barny Books. It would not have been possible without the enthusiastic help and support of the current and former Headmasters, the Staff and the Boys of The King's School, Grantham. Mr Charles Dormer has provided our Foreword, Mrs Sue Long, Assistant Headteacher and Business and Enterprise Specialism Director, was the Project Manager, Mrs Linda Dawes and Mrs Jo Snee, School Librarians, supported the Illustrators and Writers throughout. Mrs Alison Cherry was our link with the Mathematics Department and St Wulfram's. Mrs Ruth Crook, Chairman of the PTFA and local historian, helped with details of Newton's early life and proof-read the text as did Mrs Jenny Haden.

Professor Valerie Gibson, Professor in High Energy Physics and Fellow of Trinity College, Cambridge, very generously gave her time to welcome us to the College and Mr Sandy Paul kindly showed us Newton's books in the Trinity College Wren Library.

At Woolsthorpe, Mr Stuart Crow of the National Trust kindly agreed to our week's workshop, expertly led by Ms Kim Barnett with the help of the Volunteers, Miss Vivienne Orr and Mr Alan Lievesley with Mrs Margaret Winn providing her Woolsthorpe area knowledge. Miss Sarah Brown, The King's School fund-raiser, supported us and the printing team from GB Winstonmead turned the text into the book. The cover picture of Sir Isaac Newton is from a silk-screen print given to The King's School by the NWNU High School, Lanzhou, Gansu Province, China to mark the link between the schools and to celebrate the life and work of the Grantham boy who became the world's greatest mathematician and scientist.

Sources

We found the following books and web-sites particularly useful in researching this story:

Ackroyd, Peter *'Newton'* Vintage 2006
Atkinson, Mary *'Genius or Madman? Sir Isaac Newton'* Scholastic 2008
Gleick, James *'Isaac Newton'* Harper Perennial 2003
Hawking, Stephen *'On the Shoulders of Giants: The Great Works of Physics and Astronomy'* Running Press 2002
Latham, Robert (Ed) *'The Shorter Pepys'* Penguin Books 1985
Newton, Sir Isaac and Motte, Andrew *'Principia'* Great Minds Series University of California Press 1995
Poskitt, Kjartan *'Isaac Newton and his Apple'* Scholastic Children's Books 1999
Westfall, Richard S *'Never at Rest, a Biography of Isaac Newton'* Cambridge University Press 1983
White, Michael *'Isaac Newton: The Last Sorcerer'* Fourth Estate 1997

www.isaacnewtonsoldschool.org/
www.kings.lincs.sch.uk/
www.nationaltrust.org.uk/main/w.../w-woolsthorpemanor/
www.newton.ac.uk/newtlife.html
www.newton.ac.uk/newton.html
www.newtonproject.sussex.ac.uk/
royalsociety.org/
350.royalsociety.org/
www.trin.cam.ac.uk/

and many others

Thanks

This book is a 'Books with Schools' Project within the series published by Barny Books. It would not have been possible without the enthusiastic help and support of the current and former Headmasters, the Staff and the Boys of The King's School, Grantham. Mr Charles Dormer has provided our Foreword, Mrs Sue Long, Assistant Headteacher and Business and Enterprise Specialism Director, was the Project Manager, Mrs Linda Dawes and Mrs Jo Snee, School Librarians, supported the Illustrators and Writers throughout. Mrs Alison Cherry was our link with the Mathematics Department and St Wulfram's. Mrs Ruth Crook, Chairman of the PTFA and local historian, helped with details of Newton's early life and proof-read the text as did Mrs Jenny Haden.

Professor Valerie Gibson, Professor in High Energy Physics and Fellow of Trinity College, Cambridge, very generously gave her time to welcome us to the College and Mr Sandy Paul kindly showed us Newton's books in the Trinity College Wren Library.

At Woolsthorpe, Mr Stuart Crow of the National Trust kindly agreed to our week's workshop, expertly led by Ms Kim Barnett with the help of the Volunteers, Miss Vivienne Orr and Mr Alan Lievesley with Mrs Margaret Winn providing her Woolsthorpe area knowledge. Miss Sarah Brown, The King's School fund-raiser, supported us and the printing team from GB Winstonmead turned the text into the book. The cover picture of Sir Isaac Newton is from a silk-screen print given to The King's School by the NWNU High School, Lanzhou, Gansu Province, China to mark the link between the schools and to celebrate the life and work of the Grantham boy who became the world's greatest mathematician and scientist.